12/05

WITHDRAWN

Reading Essentials
in Science

ENERGY WORKS!

Heat

JENNY KARPELENIA

PERFECTION LEARNING®

Editorial Director:	Susan C. Thies
Editor:	Mary L. Bush
Design Director:	Randy Messer
Book Design:	Mark Hagenberg
Cover Design:	Michael A. Aspengren
Photo Research:	Lisa Lorimor

A special thanks to the following for their scientific review of the book:

Judy Beck, Ph.D.; Associate Professor, Science Education; University of South Carolina-Spartanburg

Jeffrey Bush; Field Engineer; Vessco, Inc.

Image Credits
© Bettmann/CORBIS: pp. 6 (right), 36 (bottom); © Patrick Giardino/CORBIS: p. 8 (bottom);
© LWA-Sharie Kennedy/CORBIS: p. 10

ArtToday (some images copyright arttoday.com): cover (left; center), pp. 1 (left; center), 4, 5, 6 (left), 9,
14, 17, 18 (top background), 19, 22, 25, 26, 27, 32 (top), 34 (bottom), 37; Corel: cover (right), pp. 1 (right),
2–3, 11 (top right), 18 (bottom), 23, 28, 30 (top left), 36 (top); Corbis: pp. 12, 24, 35;
Dynamic Graphics: cover (background), pp. 11 (top left), 13, 16, 21, 29; Digital Stock: pp. 30 (top right),
31, 32 (bottom), 33; Photodisc: pp. 7, 8 (top); Mike Aspengren: p. 34 (top); Wade Thompson: p. 11 (bottom)

For information, contact
Perfection Learning® Corporation
1000 North Second Avenue, P.O. Box 500
Logan, Iowa 51546-0500.
Phone: 1-800-831-4190
Fax: 1-800-543-2745
perfectionlearning.com

3 4 5 6 7 8 PP 09 08 07 06 05 04

Paperback ISBN 0-7891-6060-9
Reinforced Library Binding ISBN 0-7569-4449-X

Contents

Introduction to Energy

ENERGY—WHAT IS IT?

Was there ever a time when you felt so tired that you could not even shoot some hoops or play your favorite video game? Maybe you were feeling sick or very hungry. You probably felt as if you had no **energy**.

Now think of a time when you had lots of energy. You felt as if you could run, talk, ride your bike, or play games forever. Perhaps your parents or teachers even told you that you had *too much* energy.

So what is energy? What forms of energy are there?

Energy is the ability to get things done or to do work of some sort. Anything that accomplishes something is using some form of energy. When you hear the word *work*, do you think of chores around the house? *Work* actually means getting *anything* done. A football sailing through the air is using and giving off energy. A ringing doorbell is using and giving off energy. A shining lightbulb is using and giving off energy.

These examples also show some of the different forms of energy. A thrown football is an example of motion energy. A ringing doorbell is using electrical energy and giving off sound energy. A lightbulb uses electrical energy and gives off light and heat energy. Motion, electricity, sound, light, and heat are all forms of energy. These forms of energy affect our lives every day.

A streetlamp uses electricity to produce both light and heat energy.

Carnival rides use electrical energy to create motion energy.

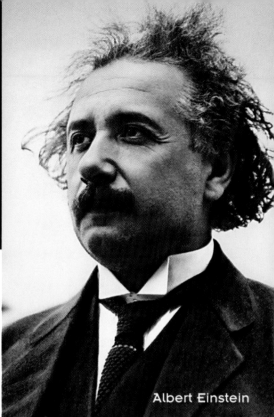

Albert Einstein

IT'S A LAW

Scientists perform experiments to test their **theories**, or ideas, about the world. Experiments that produce the same results over and over become scientific laws. One of these laws says that energy cannot be created or destroyed, but it can change from one form to another. This means that all around us, every day, energy is being changed from one form to another. The amount of energy in the universe stays the same, but it is constantly taking different forms.

The famous scientist Albert Einstein was a great thinker. He thought of new ideas that other people had not even imagined. He developed the equation $E = mc^2$. The E stands for "energy." The m stands for "mass" (the amount of "stuff" in an object). His idea shows that energy and mass can change back and forth. So energy can be changed into stuff, and stuff can be changed into energy.

ENERGY WORKS!

Energy is very important. It allows many types of work to be done. People have energy. Plants have energy. The Sun gives off energy. Machines use energy. Read on to find out more about energy, its forms, how it works, and how it is used.

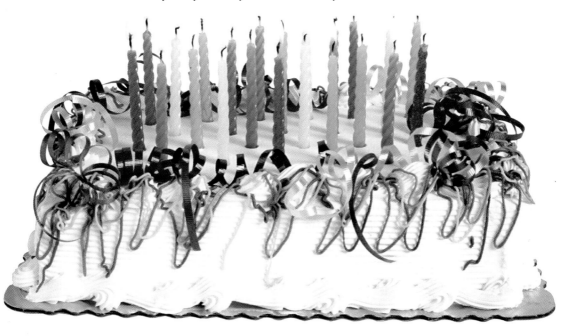

CHAPTER 1

Hot Stuff

You and your friends are munching on pizza and watching a movie. Wrapping paper and boxes are scattered all over. Your new presents lie in a pile. It's your birthday.

Your mom brings in the cake she baked for you. She lights the candles while you make a wish. You pass around plates piled with cake and melting ice cream.

When the day is over, your friends head out into the chilly air. You sort through your presents, wishing every day was your birthday.

But did you notice the extra guest moving about at your party? This guest was there when your mom put the cake in the oven. The guest helped bake the pizza you and your friends enjoyed. The guest was there when the candles on your cake were lit. The guest hung around as the ice cream melted. The guest was there for the whole party, keeping your friends and family warm inside your home. Who or what was this extra guest? It was **heat**.

Heat energy surrounds us every day. The Sun gives our planet heat. Food provides our bodies with heat energy. Heat warms our homes and helps make electricity for our TVs, stereos, and video games. It would be a cold, dull world without this important form of energy.

CHAPTER 2

You're Getting Warmer

HOT AND COLD

What is heat? Heat is energy given off by moving **molecules**. Molecules are small particles that make up all the things around us. Molecules make up the air, soil, and water. Your desk, your bike, and even *you* are made up of molecules. These molecules are always moving, creating heat energy.

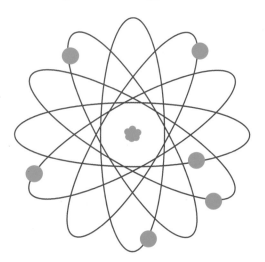

Another Name for Heat Energy

Heat energy is also called *thermal energy*. *Therm* or *thermo* means "heat." You might wear thermal underwear in the winter to stay warm. You can put hot chocolate in a thermos to keep it warm. You use a thermometer to measure **temperature**. A thermostat controls the heating and cooling in your home. All of these things are related to heat, or thermal, energy.

What is **cold**? Cold is not a form of energy. Cold is just a loss of heat. When heat energy leaves an object, it feels colder. The food you put in a refrigerator stays cold because the heat is removed from the refrigerator. When you exercise and sweat, the water on your skin takes heat away from your body. That cools you off.

Put some hot water in a bowl. Put some cold water in another bowl. Measure and record the temperature of the water in each bowl with a thermometer.

Now mix an equal amount of the hot and cold water together in a third bowl. Measure and record the temperature. How does the mixed water temperature compare to the separate water temperatures?

Record the temperatures of all three bowls every two minutes. How long does it take for the bowls of water to reach the same temperature? Why does this happen?

Heat always moves from hotter to cooler areas. So the hot water will lose heat and the cold water will gain heat until the temperatures are equal.

Molten lava from an erupting volcano

Greenhouse

SOURCES OF HEAT

Where does heat come from? There are several sources of this form of energy.

The Sun is the most important source of heat on Earth. All life on the planet depends on the Sun's heat. Light and heat energy from the Sun is called *solar energy*.

The Earth also has heat deep inside. Volcanoes and **geysers** release some of this heat. Heat energy from inside the Earth can be used to make electricity. It can also provide heat for small areas such as **greenhouses**.

Chemical reactions occur when two or more substances are mixed together and produce a new substance. During most chemical reactions, heat energy is released. Fire is a chemical reaction caused when a fuel mixes with oxygen.

Fire is a very common source of heat energy. When objects burn, they produce heat. This energy can be used to heat buildings and cook food. Fire is also used to make tools and other products in factories.

Steam rises from a geyser at Yellowstone National Park.

Try This!

Rub your hands together. What do you notice? You should feel your hands getting warmer. When objects rub against each other, heat is created.

Gymnasts rub chalk powder on their hands to reduce the friction between their hands and the bars.

Friction is a **force** that acts against motion. When two objects rub together, they each push or pull against the motion of the other. This creates heat. Large amounts of friction can produce fire, such as when two sticks are rubbed together to make a campfire.

Nuclear energy is another source of heat. All substances are made up of tiny particles, or atoms. When the center, or nucleus, of certain atoms splits apart, heat energy is released. This heat can be used to make electricity.

What's the Difference?

Atoms are the smallest particles that make up all matter, or stuff, in the world. *Molecules* are one or more atoms grouped together.

CHAPTER 3

Heat Changes Things

Heat energy causes changes in substances. When heat is added or lost, objects can change form, change into new substances, or change size.

PHYSICAL CHANGES

Adding heat energy to a substance can change its form, or how it looks. A solid has its molecules arranged in a neat, tight pattern. This gives a solid its particular shape and size. Heating a solid can melt it into a liquid. The extra heat makes the solid's molecules move faster. The molecules can slide past one another quicker and easier. They are able to flow and change shape.

When the object has changed to a liquid, it has a new form. It is still the same substance, and there is still the same amount of it. When an ice cube melts, the same amount of water is in the liquid puddle as there was in the solid cube.

Adding heat to a liquid can change it into a gas. This process is called *evaporation*. The added heat makes the liquid's molecules move even faster. The molecules move about freely, far apart from one another. A gas is able to change shape and size as the molecules constantly move around and spread apart.

Heating a liquid at high temperatures until it changes to a gas, or steam, is called *boiling*. This is what happens when you cook water on the stove and bubbles of gas form and rise into the air.

Liquids can also evaporate into gases on their own. A puddle dries up outside in the warm air. The water in the puddle is actually evaporating into the air. If you spill water on the floor, eventually it will evaporate—hopefully before someone steps in it!

Skipping a Step

When heat is added, most substances change from a solid to a liquid to a gas. But a few substances change from a solid right to a gas. This is called *sublimation*.

Dry ice is an example of a substance that sublimates. Dry ice is carbon dioxide in a solid form. When taken out of the freezer, it changes directly to a gas.

Try This!

Fill a small jar with dried beans. Shake the jar gently. What happens to the beans? This is how molecules act in a solid. They move slightly but stay in the same place.

Now close the jar with the lid and shake a bit harder. What happens to the beans? Do they start to move around in the jar? This is what happens when a solid changes to a liquid. The molecules move around more freely.

Remove the lid and shake hard. What happens? Did some of the beans jump out of the jar? This is what happens when a liquid turns into a gas. The molecules "jump" out of the liquid as they change into a gas.

Removing heat energy from substances also changes their form. When a gas cools, the molecules slow down and move closer together to form a liquid. This is called *condensation*. When steam in a boiling pot hits the lid and cools, it turns back into a liquid. This is why water droplets form on the lid. When water vapor, or gas, in the air cools and condenses, it becomes rain.

But what happens when air temperatures continue to get colder? Eventually, the rain will become snow or ice. Taking away heat from a liquid will cause the molecules to slow down even more and move even closer together. The liquid will then freeze, or form a solid.

Melting, boiling, evaporating, condensing, and freezing are all **physical changes**. The substance takes a different form on the outside but is still made up of the same molecules inside.

Physical Changes

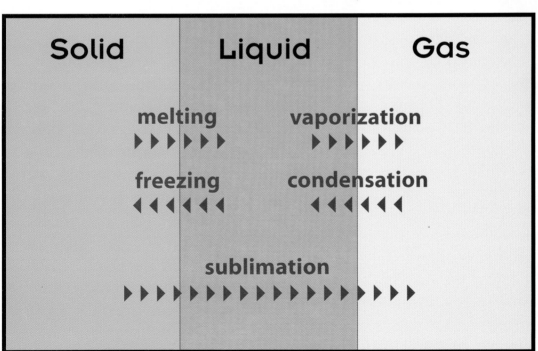

Solid	Liquid	Gas
melting ▶▶▶▶▶▶	vaporization ▶▶▶▶▶▶	
freezing ◀◀◀◀◀◀	condensation ◀◀◀◀◀◀	
sublimation ▶▶▶▶▶▶▶▶▶▶▶▶▶▶▶▶		

Note: Heat is added when moving from left to right.
Heat is removed when moving from right to left.

Frying an egg is a chemical change. The heat from the stove changes the molecules in the egg as it cooks. Think about the difference between a raw and a cooked egg. That's the power of heat!

CHEMICAL CHANGES

Adding heat to substances can also cause **chemical changes**. Chemical changes occur when heat causes an object's molecules to change into different molecules. Both the outside *and* inside of the substance is now different.

Baking a cake is an example of a chemical change. The heat inside the oven causes a chemical reaction among the ingredients, changing them into a cake. Leaving the cake in the oven too long can cause another chemical reaction. Burning is a chemical change that will turn the cake into a hard black lump.

Helpful Hint

Another way to remember the difference between physical and chemical change is to think about whether or not you can change the substance back into the same thing. In a physical change, the substance can change back and forth over and over, and it will still be the same thing. Water that boils and then condenses and then freezes is still water. In a chemical change, the substance cannot be changed back and forth. Once the egg is cooked, it can never go back to its raw state. It is now a new substance.

The heat in an oven causes a chemical change when baking cookies.

BIGGER OR SMALLER

Heat can also change the size of an object. Adding heat makes most substances expand, or grow larger. The heat makes the molecules of the substance move faster and bump into one another more. The molecules spread out farther, so the substance takes up more room than it did before.

Removing heat makes most substances contract, or get smaller. Removing heat makes the molecules of the substance move slower. The molecules crowd together, so the substance takes up less room than it did before.

Air and other gases expand and contract when heated and cooled. Have you ever noticed how a boiling pot can cause the lid to shake or push upward? That's because the steam takes up more room as it gets hotter and it needs more room to spread out. If you turn the heat off, the gas will contract and stop pushing on the lid.

Heat a marshmallow in the microwave for about ten seconds. Watch it carefully. What happens to the marshmallow?

Marshmallows are mostly sugar and air. When the microwave heats up the marshmallow, the air inside expands. This makes the marshmallow puff up.

Most liquids also expand when heated and contract when cooled. During the summer, a car's gas tank can hold less gasoline than it can in the winter. This is because warmer temperatures make the gas expand, so less can fit in the tank. Cold temperatures cause the gasoline to contract, so more can fit in the tank.

What About Water?

If you've ever left a bottle of water or a can of pop (which is mostly water) in freezing temperatures for too long, then you know that water does not contract as it gets colder. It does just the opposite. Water actually expands when it freezes. Ice takes up more room than water, which is why your bottle or can bulged or exploded when frozen. This is because of the unique way that the molecules in water are arranged.

Most solids also expand in the heat and contract in the cold. On hot summer days, concrete roads heat up. Without room to expand, the concrete is forced upward. The road buckles and breaks. Sidewalks are now poured in slabs, or sheets. The gap between each slab allows for the concrete to expand in hot weather. This keeps sidewalks from buckling and breaking.

Train tracks are designed with special joints, or connections, between rails. These joints allow for expansion and contraction as the temperature changes. The tracks are safer and last longer.

Ice skaters glide across frozen water.

CHAPTER 4

Ouch!
That's Hot!

It's your turn to do the dishes. You fill up the sink with hot water and detergent. You put your hands in. Ouch! It's way too hot. Then your mom walks by and sticks her hand in the water, searching for the dishrag. The water feels just right to her. Why is there a difference?

Fill three bowls—one with hot water, one with ice water, and one with lukewarm water. Put one hand in the hot water and one in the cold water. Keep them there for a minute or two. Then quickly put both hands in the lukewarm water. How do your hands feel?

The lukewarm water should feel very different to the two hands. Why? The hand that was in the cold water quickly absorbed the heat in the warm water, making your hand feel hot. The hand from the hot water lost heat to the warm water, so it felt colder.

No matter how either hand felt, the warm water was the same temperature for both hands. So sometimes, even your own hand can't tell what the real temperature is.

TEMPERATURE

A feeling of hot or cold is different for everyone. In science, however, temperature can be measured accurately. Temperature is a measurement of how hot or cold something is. It is a measure of how fast the molecules are moving in an object. Faster molecules mean a higher temperature. Slower molecules mean a lower temperature.

Temperature is measured in degrees. Most people in the United States measure temperature using the Fahrenheit scale. The Celsius scale is commonly used to measure temperature in other parts of the world.

Two Temperature Men

Daniel Fahrenheit from Germany and Anders Celsius from Sweden developed the temperature scales used most often today. Fahrenheit put mercury inside a glass tube and marked 180 degrees between the freezing and boiling points of water. His scale goes from 32 to 212 degrees.

Celsius marked 100 degrees between the freezing and boiling points of water. His scale goes from 0 to 100 degrees.

Temperatures to Know

Water boils	212°F	100°C
Average human body temperature	98.6°F	37°C
Water freezes	32°F	0°C
Absolute zero (the lowest temperature that exists)	-460°F	-273°C

Fahrenheit　　**Celsius**

THERMOMETERS

A thermometer is a tool used to measure temperature. Your home has many different types of thermometers. You probably have an oven thermometer, a refrigerator thermometer, and a fever thermometer. You might have a meat or candy thermometer used for cooking. The temperature in your house is controlled by a thermometer connected to your furnace.

Many thermometers have a substance in them that expands and contracts with heat changes. As the substance expands and contracts, it lines up with a number that represents the temperature. Mercury and ethyl alcohol are two liquids used in household thermometers.

Mercury Alert!

Mercury is a silver liquid metal used in many thermometers. Mercury can be dangerous if it soaks into the body. It can poison the blood and cause brain damage. So if your mercury thermometer breaks, never touch the mercury. Find someone to help you clean it up safely.

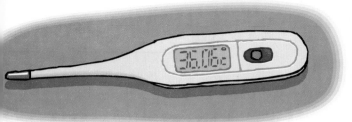

Maybe you have a digital ear or mouth thermometer at home. These thermometers work by electronically sensing heat changes inside your ear or mouth.

More About Thermometers

Check out this Web site to learn more about the types of thermometers and how they work. **http://home.howstuff works.com/therm4.htm**

The First Thermometers

In the 1600s, several scientists worked on measuring temperature. Italian scientists Galileo and Ferdinand II invented tools to measure temperature. Both inventions contained red wine inside a glass tube. English scientists Robert Hooke and Sir Isaac Newton improved the thermometer in the mid-1600s. They added markings on the side to show the freezing and boiling points of water.

Isaac Newton

Galileo

CHAPTER 5
Radiation

Heat from the Sun travels 93 million miles to warm you up. Did you ever wonder how it gets here?

HEAT WAVES

The Sun is a main source of energy on Earth. Both light and heat from the Sun travel through space in the form of waves. This energy that travels in waves is known as **radiation**. Heat is also known as infrared radiation. This is because we cannot see heat with our eyes. It has less energy than the color red on the energy spectrum.

On a hot day, it's easy to see how the Sun's heat warms us. The heat waves travel through space and are absorbed, or taken in, by our bodies.

The Sun's radiation can also be used to heat homes. Windows let in the Sun's light and heat energy. Tanks or tubes of water are heated by the Sun and sent to other parts of the house to use for cooking, bathing, and doing dishes and laundry.

OTHER HOT STUFF

Other hot objects also **radiate** heat. Lamps and ovens give off heat by radiation. A fire radiates heat. Your body even radiates some of its heat. That's why it's good to snuggle up with someone when you're cold.

Some homes use radiators for heat. Hot water is pumped through pipes and into radiators around the house. The heat waves spread out through the house.

Cars also have radiators. Heat created by the moving parts in the engine is carried away by water in the radiator. The heat waves spread out in the air and cool down. This keeps the engine from overheating and breaking down.

HOW DO WE KNOW IT'S THERE?

If we can't see heat radiation with our eyes, how do we know it's there? Humans detect radiation with their skin. Skin has heat sensors that can feel the energy waves.

Infrared equipment allows people to "see" heat waves. The heat waves given off by objects show up as different colors on a thermograph, or heat picture. Thermographs have many uses. Firefighters use thermal imaging cameras to find people in burning buildings. Scientists use heat images to determine heat levels in the oceans, clouds, and outer space. Doctors use heat images to find injuries inside the body.

Whether we can see it or not, radiation is definitely important to our lives.

Solar panels collect heat energy from the Sun.

CHAPTER 6

Convection

If you have bunk beds, the top bunk is the best place to sleep in the winter. But in the summer, the bottom bunk is the place to be. Why? Hot air rises and cool air sinks. This movement of heat through air and other gases is known as **convection**.

UP AND DOWN

The heat radiated from a hot object can heat up the air around it. That heat makes the air molecules move faster. Faster-moving molecules bump into one another more and spread farther apart. The heated air becomes less dense, or crowded together. This makes the hot air lighter, so it rises.

As it moves away from the hot object, the air begins to lose heat. The air molecules slow down. Slower-moving molecules crowd closer together. The air becomes heavier and sinks. This rising and sinking action is called a *convection current*.

Flying on Hot Air

Rising columns of hot air are called *thermals*. Birds can ride on thermals to save energy. The hot air lifts them upward, so they don't need to flap their wings as often. People riding in gliders and hang gliders also ride on thermals to soar more easily.

IT WORKS IN WATER TOO

Heat moves by convection through liquids too. Warm water molecules move around faster and spread out more than cooler water molecules. So warm water rises, while cold water sinks. This forms a convection current in the water.

Water in a pan on the stove is heated this way. The water on the bottom of the pan heats up first since it's closer to the heat source. As the warm water rises, the colder water sinks so it's now closer to the heat source. The current keeps moving throughout the pan until all the water is the same temperature.

Find two small clear plastic or glass bottles that have the same size mouth (opening). Put two drops of blue food coloring into one bottle. Fill it to the top with cold water. Put two drops of red food coloring into the other bottle. Fill it to the top with hot water.

Hold an index card or piece of cardboard over the mouth of the cold bottle. Carefully flip the cold bottle over, keeping the top covered tightly. Match up the mouths of the two bottles. Gently slide the card out from between them.

What happens? Why? The hot water should rise, and the cold water should sink. This causes the red and blue coloring to mix and make purple water.

Now rinse out the two bottles, and fill them up the same way again. This time put the cold blue water on the bottom and the hot red water on top. What happens now? Why?

CHAPTER 7

Conduction

Have you ever left a metal spoon sitting in a pot of soup that's cooking on the stove? What happened to the spoon? Did it get so hot that you couldn't pick it up or you burned yourself?

Heat can pass from one substance to another through **conduction**. Hot, fast-moving molecules bump into one another and the molecules next to them. This makes the nearby molecules move faster and bump into one another. So if a hot object touches a cooler object, molecules from the hot object bump into molecules from the cooler object. This causes the molecules in the cooler object to speed up, making the object hotter.

Try This!

Get some help from a parent or teacher for this experiment. You will need a metal butter knife and a lighted candle.

Drip three small dabs of wax about the same size along the flat blade of the knife. Let the dabs cool. Grip the knife by the end of the handle with a hot pad or oven mitt. Hold the knife vertically just above the flame of the lighted candle. Watch the wax dabs. After a short time, you should see the wax dabs start to melt. In which order did they melt? Why?

The plastic on the end of this spoon will stop the conduction of heat up the handle. Plastic does not conduct heat like metal does.

When hot molecules from the soup bumped into molecules in the spoon, the spoon's molecules started moving faster. The heat was conducted, or passed, up the spoon by bumping molecules until the top was so hot that you couldn't touch it.

CONDUCTORS AND INSULATORS

Some materials are better at conducting heat than others. Most metals are good **conductors**. Copper and stainless steel are two metal conductors. They are often used to make pots and pans to heat food. Silver and aluminum are also good heat conductors.

An **insulator's** molecules do not pass along heat well. Air is an insulator. Wood, plastic, rubber, and feathers are also good insulators. These materials are often used to keep heat in or out.

Coolers and thermoses are made with insulating materials so heat cannot move in or out. Homes are covered with insulating material to keep heat from escaping in the winter or pouring in during the summer. Winter clothes are often insulated to prevent the loss of body heat.

This fisherman wears heavy clothing to stay warm in the cold outdoors.

CHAPTER 8
Body Heat

All humans and animals need a certain amount of body heat to survive. Being too warm or too cold is harmful to living things. Humans and animals have different ways to control their body heat.

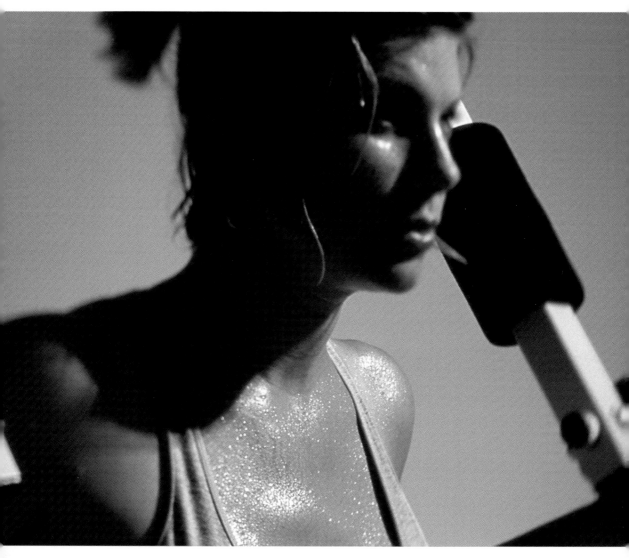

Collared lizard

Tree frog

COLD-BLOODED

Some animals are cold-blooded. This means that an animal's body temperature depends on its environment. If the weather is warm, the animal's temperature will rise. When it's cold outside, the animal's temperature will drop.

Cold-blooded animals have to move to a warmer place when it gets too cool. They have to move to a cool place when it gets too warm. For example, a turtle is cold-blooded. A turtle lays out in the Sun when it needs to warm up. When it needs to cool off, it goes into the water or shade. Other animals such as snakes, lizards, and frogs are also cold-blooded.

WARM-BLOODED

Other animals, such as birds and mammals, are warm-blooded. This means that the animal has a stable body temperature that does not depend on the environment. However, body heat can still be lost or gained, so some **adaptations** are necessary.

Human, Mammal, or Both?

A mammal is an animal that is covered with hair or fur and has the ability to feed its young with milk from the mother. Since humans fit this description, they are mammals.

Food is the main source of heat energy for warm-blooded animals. Humans and animals must eat enough food to maintain their normal body temperatures.

Different coverings help keep body temperatures constant. Humans have hair on their heads to keep heat from escaping. They also wear thick, heavy clothes in the winter to trap body heat. Light clothing in the summer lets extra heat out.

Orangutans are warm-blooded mammals that have long reddish brown hair to keep them warm.

Birds are covered with feathers to keep them warm or cool. Air spaces between feathers act as insulators. They trap the heat near the bird's body if the weather is cold. They can also keep the heat away from the bird's body if the weather is warm.

Mammals have hair or fur as a body covering. Bears, dogs, and cats are just a few mammals with furry coats. The air trapped between the hairs or fur acts as an insulator. It can keep the animal warm in the winter and cool in the summer.

Thank You, Birds

Humans also use bird feathers as insulators. Feathers are used in pillows, blankets, and winter coats to help control body temperature.

Swan

Dolphin

Some mammals, such as dolphins and whales, have much less hair. These mammals have body fat instead. Sea mammals have a thick layer of blubber, or fat, under their skin. This blubber acts as an insulator to keep their temperatures constant.

Special body functions can also help control body temperature. Have you ever gotten the chills after getting out of the water from a pool, bath, or shower? Evaporating water pulls heat away from the body and into the air. Sweating helps do the same thing. Humans sweat to lower their body temperature. It is a built-in method to keep them from getting too hot.

Some mammals cannot sweat.

Pigs roll in the mud to keep cool. The mud acts as an insulator against heat absorbtion. Dogs pant. Water evaporates off a dog's tongue, pulling heat with it. Cats stretch out to give off more heat. Elephants flap their ears to release heat into the surrounding air.

Squirrels with Antifreeze?

Some animals, such as squirrels, have a special substance in their blood. It keeps squirrels' blood from freezing when they hibernate through the winter. This chemical is similar to the antifreeze used in cars to prevent the engine from freezing.

CHAPTER 9

Helpful Heat

Have you ever taken the time to think about how helpful heat is? Heat from the Sun warms us. It provides the warmth that plants and animals need to live. Heat cooks our food and warms our homes. Heat energy can also be used to do work for us. Many machines use heat to perform jobs that make our lives easier and more comfortable.

Counting Calories

Heat can be used to figure out how much energy is in foods. Scientists can burn a food sample and measure how much heat it gives off. They measure this heat in a unit called *calories*.

ELECTRICITY

Heat energy is often used to make electricity. Most power plants use a fuel to heat water to boiling. The boiling water creates steam. The steam pushes against large turbines that look like big fan blades. The turbines are connected to generators that turn the motion energy into electrical energy. The power plants then send this electricity to homes, schools, and businesses.

This power plant burns coal to produce electricity.

ENGINES

Heat energy is also used in the engines that run most of the machines we use. Engines have changed greatly over time, from the first steam engine to the combustion engine of today.

A simple steam engine was built about 2000 years ago by a Greek scientist named Hero. He used fire to heat water to boiling. The steam caused a tube to spin. Hero didn't find many useful jobs for his "engine," but his ideas were used by many others after him.

Steam train

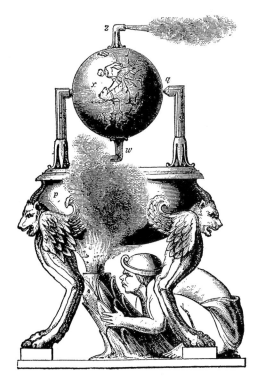

Hero's steam engine

In the late 1600s through the 1800s, steam engines became useful. Early steam engines were used to pump water from mines. In 1764, James Watt improved the steam engine by finding a way to change up-and-down motion into rotating motion. This rotation could be used to move belts and wheels in factories and mills. In 1814, George Stephenson built the first steam train. Learning how to change heat into motion energy changed industry and transportation forever.

Transportation changed even more with the invention of the combustion engine. *Combustion* means "burning." Most modern vehicles use combustion engines. A fuel, such as gasoline, is mixed with the oxygen in the air and burned inside the engine. The mini-explosions from burning the gasoline force the pistons to move up and down. The pistons are connected to a crankshaft that turns the wheels of the vehicle. Heat energy is safely changed into motion energy to get cars, trucks, tractors, and lawnmowers moving.

THINK ABOUT IT!

Most people take heat for granted. They don't think about body heat, or heat from the Sun, or even where the heat in their homes comes from. But imagine life without this warm form of energy. Isn't that a chilling thought?

Car engine

INTERNET CONNECTIONS AND RELATED READING FOR HEAT

Energy Quest
(**http://www.energyquest.ca.gov/index.html**)
This fun site provides information, stories, news, projects, games, and links to other energy sites. Visit the Gallery of Energy Pioneers too.

How Stuff Works
(**http://www.howstuffworks.com**)
If you have questions about how anything involving energy (and anything else!) works, this Web site is the place to look. It includes sections on energy and electrical power, light, sound, heat, and many inventions related to energy.

Energy Information Administration
(**http://www.eia.doe.gov/kids/index.html**)
Review the definition of energy and its forms here. Then check out the Kid's Corner, Fun Facts, and Energy Quiz.

U.S. Department of Energy
(**http://www.eere.energy.gov/kids/**)
Dr. E's Energy Lab will teach you about solar energy and energy efficiency. A dog named Roofus shows you his energy-efficient home and neighborhood. Many links to other energy sites can be found here.

The Atoms Family
(**http://www.miamisci.org/af/sln**)
This spooky Web site teaches about different forms of energy through simple experiments.

One World.Net's Kid's Channel
(**http://www.oneworld.net/penguin/energy/energy.html**)
Tiki the Penguin discusses the positive and negative sides of different types of energy sources.

Day Light, Night Light: Where Light Comes From by Franklyn M. Branley. Discusses the properties of light, particularly its source in heat. HarperCollins, 1998. [RL 2.3 IL 2–4] (5666601 PB 5666602 CC)

Energy by Jack Challoner. An Eyewitness Science book on energy. Dorling Kindersley, 1993. [RL 7.9 IL 3–8] (5868606 HB)

Energy by Alvin and Virginia Silverstein and Laura Silverstein Nunn. Explains a fundamental concept of science, gives some background, and discusses current applications and developments. Millbrook Press, 1998. [RL 5 IL 5–8] (3111906 HB)

Heat by Darlene Lauw and Lim Cheng Puay. Simple text and experiments describe and demonstrate the principles of heat and how heat energy is produced. Crabtree Publishing, 2002. [RL 3.5 IL 2–5] (3396501 PB)

• RL = Reading Level   • IL = Interest Level

Perfection Learning's catalog numbers are included for your ordering convenience. PB indicates paperback. CC indicates Cover Craft. HB indicates hardback.

GLOSSARY

adaptation (ad ap TAY shuhn) adjustment to conditions in the environment

chemical change (KEM uh kuhl chaynj) change in a substance's molecules; change from one substance to another

cold (kohld) loss of heat energy

conduction (cuhn DUHK shuhn) movement of heat from one material to another

conductor (cuhn DUHK ter) material that allows heat to move through it easily

convection (cuhn VEK shuhn) movement of rising hot air and sinking cold air

energy (EN er gee) ability to do work

force (fors) push or pull

geyser (GEYE zer) rush of hot water and steam that rises from beneath the earth

greenhouse (GREEN hows) building used to grow plants

heat (heet) energy given off by moving molecules

insulator (IN suh lay ter) material that doesn't allow heat to move through it easily

molecule (MAH luh kyoul) small particle that all substances are made of

physical change (FIZ uh kuhl chaynj) change in a substance's form

radiate (RAY dee ayt) to give off or spread (heat waves)

radiation (ray dee AY shuhn) energy that travels in the form of waves

temperature (TEM puh cher) measure of how hot or cold an object is

theory (THEAR ee) belief that has been scientifically tested to the point of being accepted as true by most people

INDEX